Learn with Biplob

Verbs

Have	Has

Q. Complete the sentences with correct form of verb

1. She ______________ a flower.

2. Aditya ______________ done this origami.

3. I ______________ some honey.

4. I ______________ come to Biplob's garden.

5. We ______________ taken Floppy the dog for a walk.

6. The bee __________ 6 legs.

7. Farmers ______________ an important job.

8. Biplob ______________ a beautiful garden.

9. Farmer Balram ______________ a lovely field of crop.

10. Spot the leopard ______________ a spotted coat.

For more sheets visit:
www.biplobworld.com

Learn with Biplob

Opposite Words (Adjectives)

Draw a line to match each word to its opposite

Sharp ●	● Big
Difficult ●	● Clean
True ●	● Low
Dirty ●	● False
Full ●	● Easy
Small ●	● Bright
High ●	● Dull
Dark ●	● Empty

Learn with Biplob

Consonant Blends

Fill in the blanks with the missing letters

fr or fl

❄️	___ ozen
🚩	___ ag
🍉	___ uit
🐸	___ og
🌸	___ ower
🪰	___ y

gr or gl

🌍	___ obe
🧤	___ ove
🍇	___ ape
🏪	___ ocery
	___ ass
🌱	___ ass

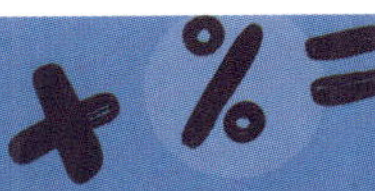

Learn with Biplob

Write the correct word the jumbled letters

lofewr ___________

piblob ___________

eloaprd ___________

armfer ___________

oobk ___________

Save Energy !

For more sheets visit:
www.biplobworld.com

Learn with Biplob

Complete the puzzle

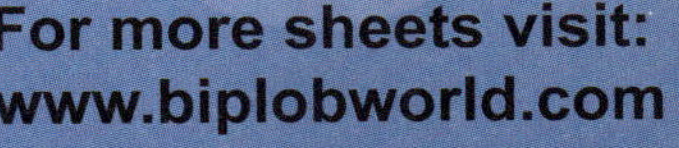

Answer
ACROSS: 4. steamship 5. car 6. bicycle 7. airship 10. airplane 11. scooter
DOWN: 1. yacht 2. helicopter 3. spaceship 6. bus 8. pickup 9. balloon

For more sheets visit:
www.biplobworld.com

Learn with Biplob

Nouns

Place the words below into the correct columns

village   farmer   sickle   car

boy   garden   forest   girl

paper   leaf   pencil   gardner

Person	Place	Thing

For more sheets visit:
www.biplobworld.com

Learn with Biplob

Adjectives include words that describe what a noun looks like and what it feels like to touch, taste or smell

Q. Find the adjectives and circle it

books	ball	yummy	slow
long	small	old	girl
mouse	car	cat	circle
fast	tall	loud	tiny
delicious	cool	home	box
large	phone	salty	cake
speedy	low	dog	baby

Same Sounds

Circle the correct homophone to complete the sentence

- Biplob's garden won a (medal / metal).

- To finish the origami, Addy stayed up all (knight / night).

- (Fir / Fur) provides Spot the leopard protection against the cold.

- This (road / rode) is too narrow for cars.

- Who wants a (peace / piece) of cake?

- This math problem doesn't make any (cents / sense) to me.

- Biplob has his breakfast at (ate / eight) o'clock in the morning.

- A (hair / hare) raced with a tortoise.

- Biplob asked farmer Balram to (wait / weight) before spraying any chemicals

- Farmer Balram started making the solar pump (right / write) away

Plant more trees !

Learn with Biplob

Compound Words

Use a word from the word box to make a compound word.
Use the pictures as clues.

sun + ___________ = ___________________

snow + ___________ = ___________________

camp + ___________ = ___________________

butter + ___________ = ___________________

jelly + ___________ = ___________________

grape + ___________ = ___________________

sky + ___________ = ___________________

Save our planet

Fact or Opinion

A **fact** is a statement which can be proven to be true
Example: Bees are very important for pollination
An **Opinion** is a personal view
Example: Addy makes the best looking origami in the world

Q. Read each sentence below and decide if it states a fact or an opinion. Circle your answer.

1. Everyone should play in the garden at least once a week   **FACT**   **OPINION**

2. Biplob the Bumblebee takes nectar from plants to make honey   **FACT**   **OPINION**

3. The energy from the sun gives us solar power   **FACT**   **OPINION**

4. Spot the Leopard runs as fast as the wind   **FACT**   **OPINION**

5. Plants need water and sunlight to grow   **FACT**   **OPINION**

Learn with Biplob

A gift shop sells gift boxes containing 20 sweets each. Here are the number of sweets sold in the week.

Each 🎁 contains 20 sweets

MONDAY	🎁	🎁					
TUESDAY	🎁	🎁	🎁	🎁	🎁		
WEDNESDAY	🎁	🎁					
THURSDAY	🎁	🎁	🎁	🎁			
FRIDAY	🎁	🎁	🎁	🎁	🎁	🎁	
SATURDAY	🎁	🎁	🎁	🎁	🎁	🎁	🎁

1. How many sweets were sold on Thursday? ________________

2. Which day were the most sweets sold? __________________

   How many? ________________

3. How many more sweets were sold on Tuesday than on Wednesday?

   ____

4. How many sweets were sold total in that week? ____________

5. There were more sweets sold on the last 2 days than on the first 4

   days? True or False? ____________________

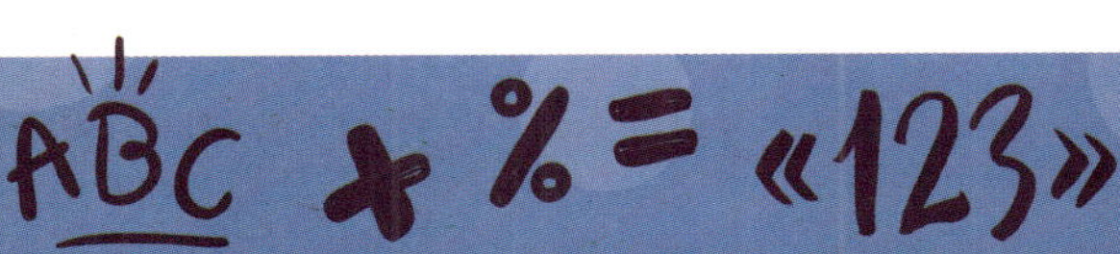

Learn with Biplob

Math Crossword

For more sheets visit:
www.biplobworld.com

Learn with Biplob

BIPLOBS CODE FOR THE WORLD

Fill in the blanks using the correct word

A	B	C	D	E	F	G	H	I	J	K	L	M
36	24	64	45	90	12	18	72	49	51	11	66	8

N	O	P	Q	R	S	T	U	V	W	X	Y	Z
86	33	79	17	44	7	80	50	39	99	16	25	61

What does Biplob say about our world?

70 + 16	6 x 6	92 - 12	30 + 20	11 x 4	62 + 28

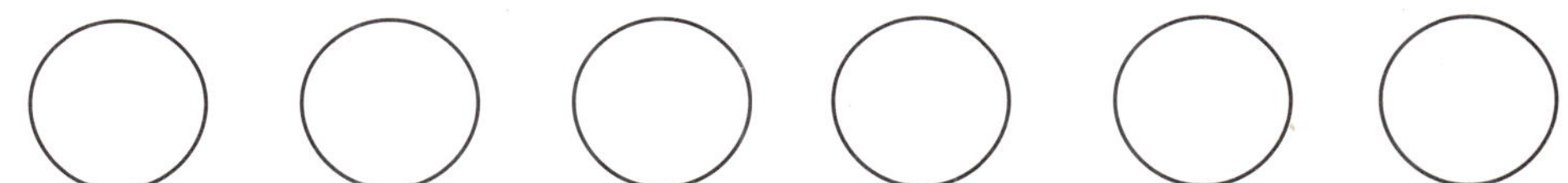

11 x 6	20 + 13	13 x 3	72 + 18	44 - 37

25 x 2	20 - 13

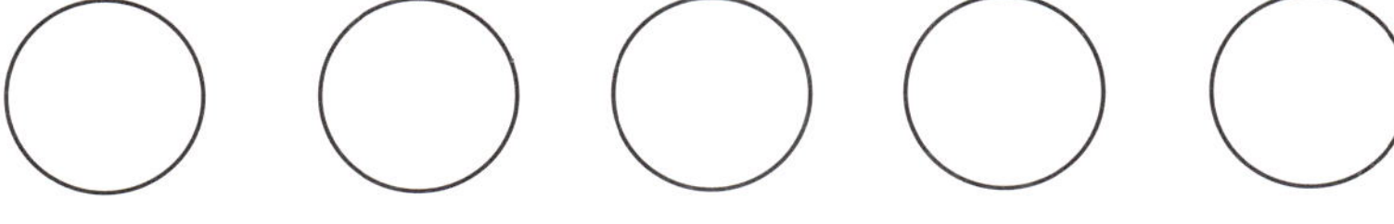
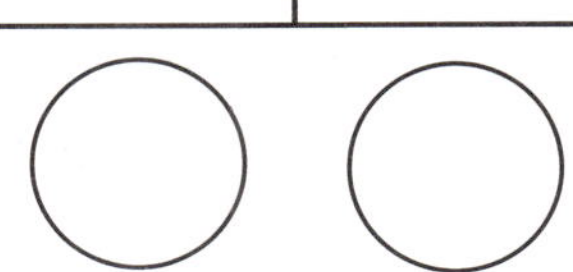

12 x 3	60 + 6	30 + 36

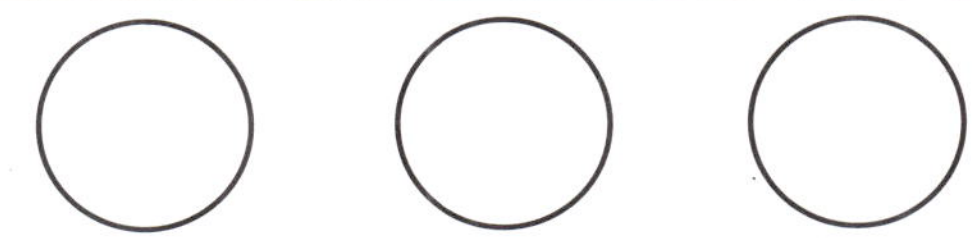

Word Problems

Q1) Yesterday Biplob's garden had 12 flowers. Today there are 9 more. How many flowers does his garden have in all? _______________________

Q2) MS. Clara the ladybug sent 8 soldiers to Biplob's garden to fight against the bad aphids. Now he has 25 ladybugs in his garden. How many ladybugs were there before MS. Clara sent 8 more?__________________

Q3) Addy made 16 origami shapes and Avantika made 9. How many more Origami did Addy make than Avantika? __________________

Q4) Biplob went to meet his sister Munmun on 3 August and returned on 21 August. How many days did Bipob spend with Munmun? ______________

Q5) When Kusaku, the leader of the langurs called them for a meeting with Biplob, only 24 of the 32 langurs could come. How many langurs were absent from the meeting? ______________________

Q6) There were 9 flowers in the garden who gave Biplob 3 spoons of nectar each. How many spoons of nectar did Biplob get in all? ________________

Q7) Spot the leopard chased 14 hunters away but 5 hunters still remained hidden in the jungle. How many hunters were there in all? _____________

Q8) Karini the elephant carried Biplob for 12 kilometers. Biplob had already flown for 7 kilometers. How many kilometers did Biplob cover in total?

____

Colour the Correct Clock

12 : 10

10 : 20

4 : 05

6 : 45

3 : 25

3 : 00

Multiply by the Center Number

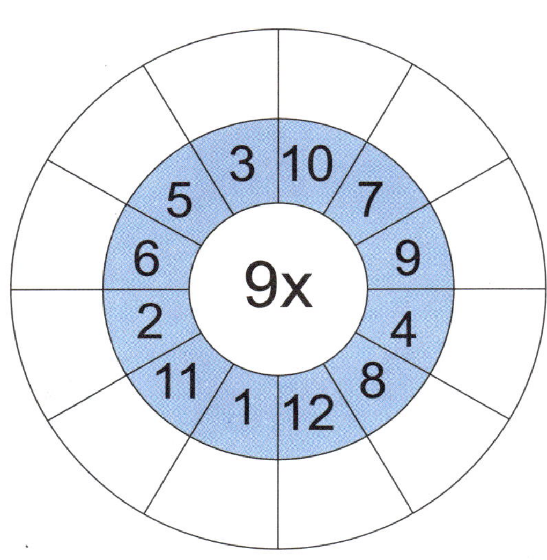

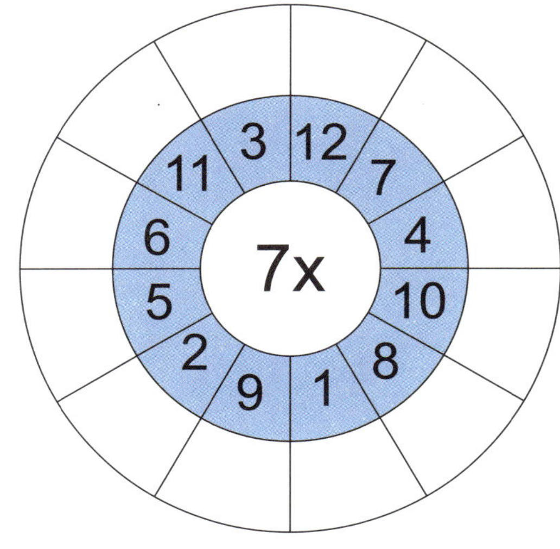

Divide by the Center Number

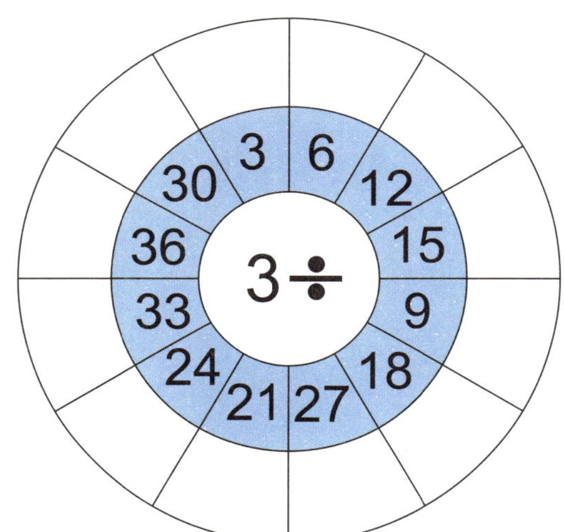

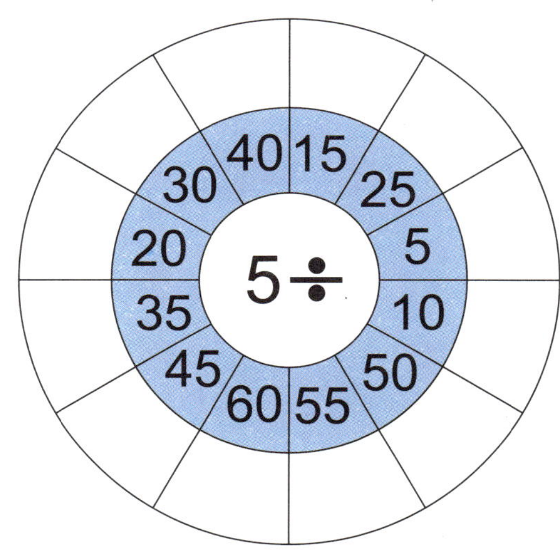

Learn with Biplob

Arrange the numbers in ascending order

179, 215, 34, 76, 5

28, 680, 480, 280, 180

156, 47, 22, 11, 633

53, 19, 75, 4, 119

Arrange the numbers in descending order

46	3	79	86	7

19	118	34	3	11

63	100	31	178	2

Save water !

For more sheets visit:
www.biplobworld.com

Learn with Biplob

Colour the Fractions

$$\frac{2}{8} + \frac{3}{8} = \frac{\square}{\square}$$

$$\frac{1}{6} + \frac{4}{6} = \frac{\square}{\square}$$

$$\frac{3}{\rule{1em}{0.4pt}} + \frac{4}{\rule{1em}{0.4pt}} = \frac{\square}{\square}$$

Plant more trees !

Learn with Biplob

The cost of each shape is given below
Add the costs to find the total cost of each picture

Rs.4 Rs.10 Rs.10 Rs.5 Rs.2 Rs.3 Rs.4

Cost: Rs. _________

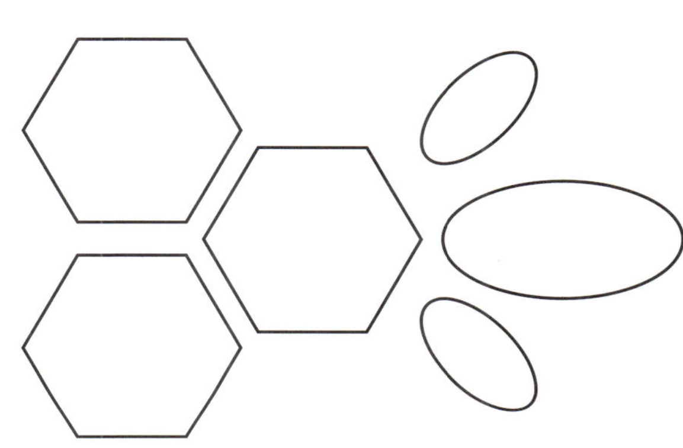

Cost: Rs. _________

Cost: Rs. _________

Eat fruits everyday

Learn with Biplob

For each of the shapes below,
1. Draw the lines of symmetry in each shape
2. Name each shape

The first one has been done for you

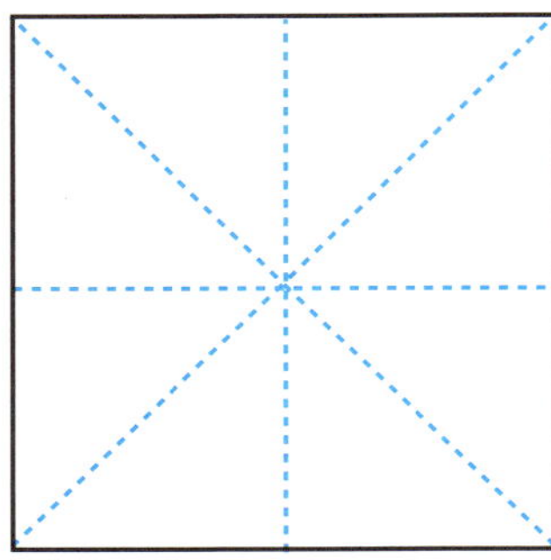

Sides: 4

Lines of symmetry: 4

Name of the shape: Square

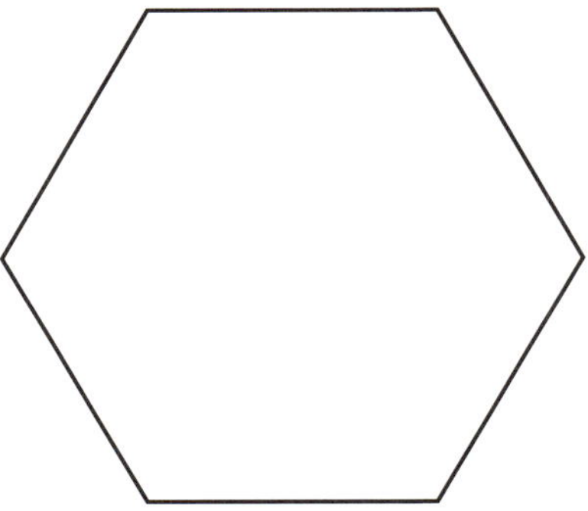

Sides:

Lines of symmetry:

Name of the shape:

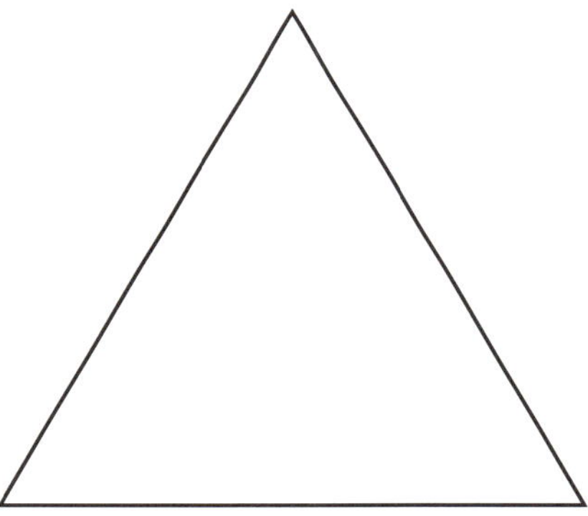

Sides:

Lines of symmetry:

Name of the shape:

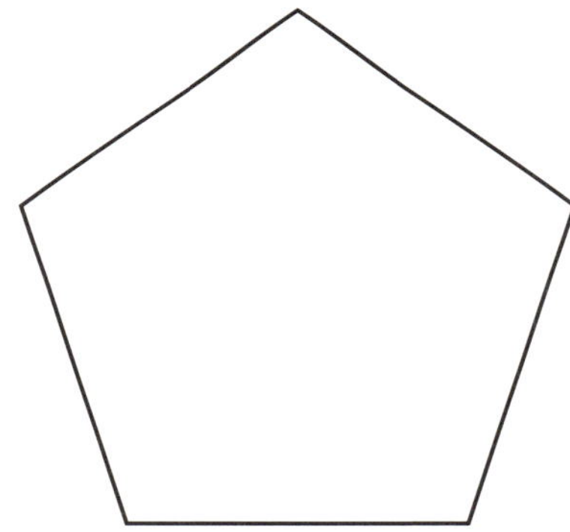

Sides:

Lines of symmetry:

Name of the shape:

Learn with Biplob

Parts of a Plant

Do you know the uses of the different parts of the plants?
Fill in the blanks from the options below and find out!

____________ **Bud**

Flower ____________

____________ **Fruit**

Leaf ____________

____________ **Stem**

Seed ____________

____________ **Root**

A. Helps in making food & reproduction
B. Supports the leaves
C. Absorbs water & Minerals
D. Prepares Food
E. Protects & helps spread the seed
F. Attracts insects for pollination
G. Helps new plants grow

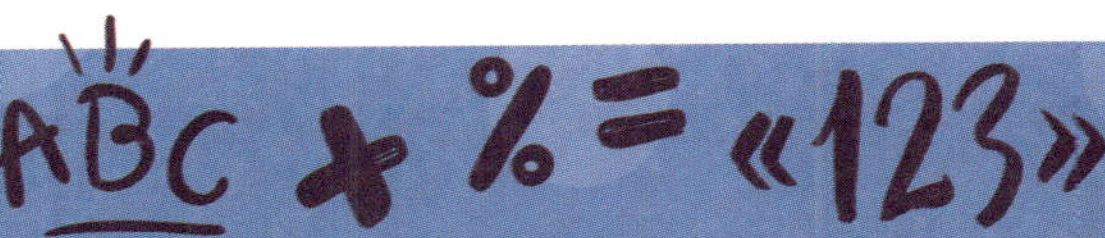

The Earth

Fill in the blanks

1. Shape of the earth is ________

2. The earth spins on it's axis. This movement is

called _________

3. The earth takes ________ hours to complete one rotation.

This is equal to ________ day

4. The earth goes around the sun. this movement

is called _________

5. Earth takes ___________to revolve around the sun. This is

equal to _______ year

6. Earth's rotation causes _______ and _________

7. Earth's revolution around the sun causes the four __________

8. The sun gives the earth _________ and __________

9. Most of the earth is covered with ___________

Learn with Biplob

Our Body

Choose from the given options, what each
part of our body is useful for. Write the correct number in the box

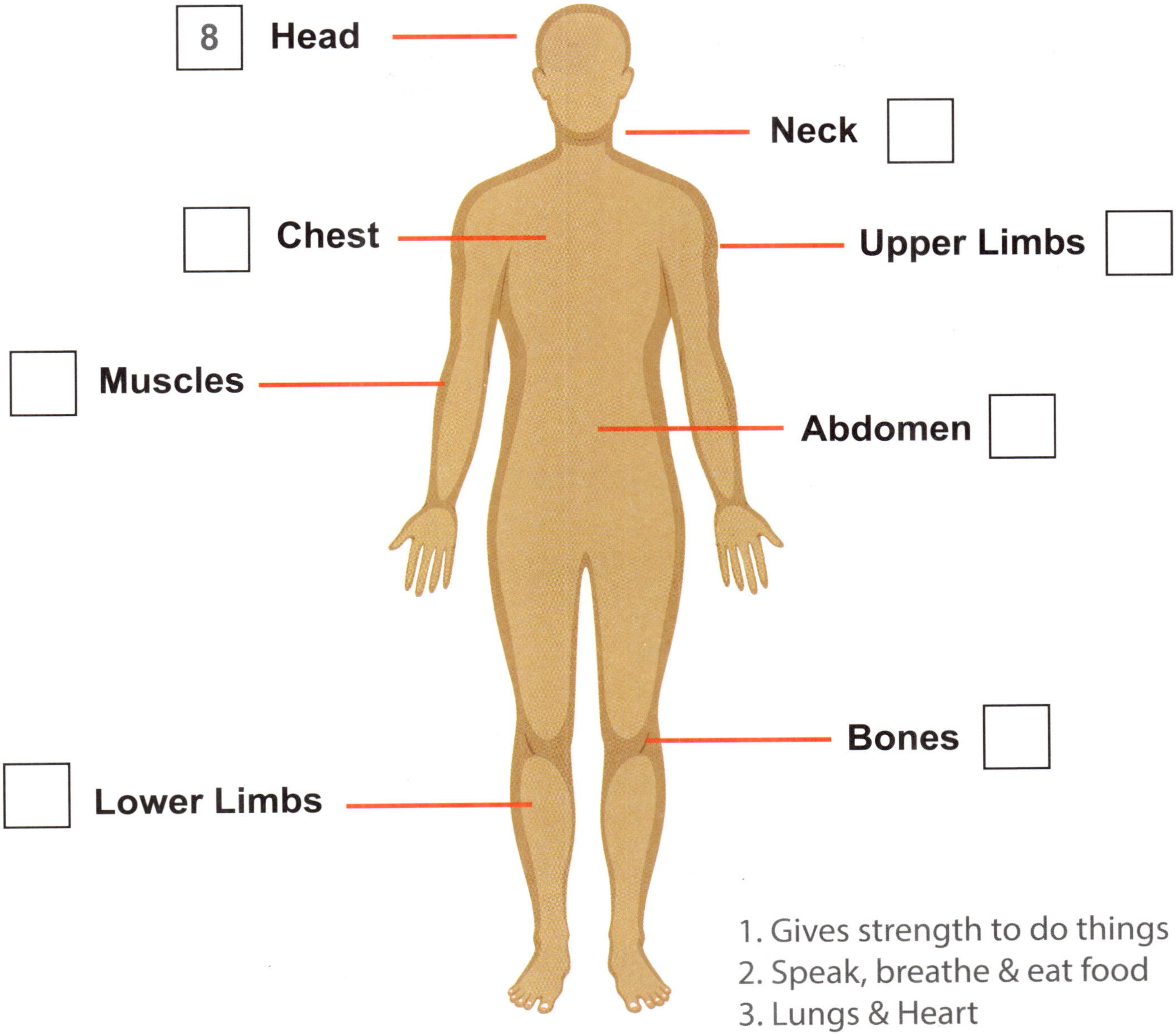

8 **Head**

Neck

Chest

Upper Limbs

Muscles

Abdomen

Bones

Lower Limbs

1. Gives strength to do things
2. Speak, breathe & eat food
3. Lungs & Heart
4. Stomach & intestines
5. Helps us do things & work
6. Helps us walk & run
7. Supports the body
8. Brain and face

For more sheets visit:
www.biplobworld.com

Learn with Biplob

Solar System

Fill in the blanks

1. All the planets revolve around the __________

2. Earth is the ____________ planet from the sun

3. The planets _________ and _________ are earth's immediate neighbours

4. ___________ is the planet closest to the sun

5. The planet farthest from the sun is __________

6. There are ___________ planets in the solar system

7. The largest planet in the solar system is ____________

8. The smallest plant in the solar system is ____________

Learn with Biplob

Sense Organs

Which sense organs do we use to sense the following.
Tick the correct organs.

THINGS	EYES	EARS	NOSE	TONGUE	SKIN
Hot Soup					
Dirty Socks					
Music					
Silk Pillow					
Pictures					

Learn with Biplob

Living or Non Living

Tick ✓ for living and **X** for non living

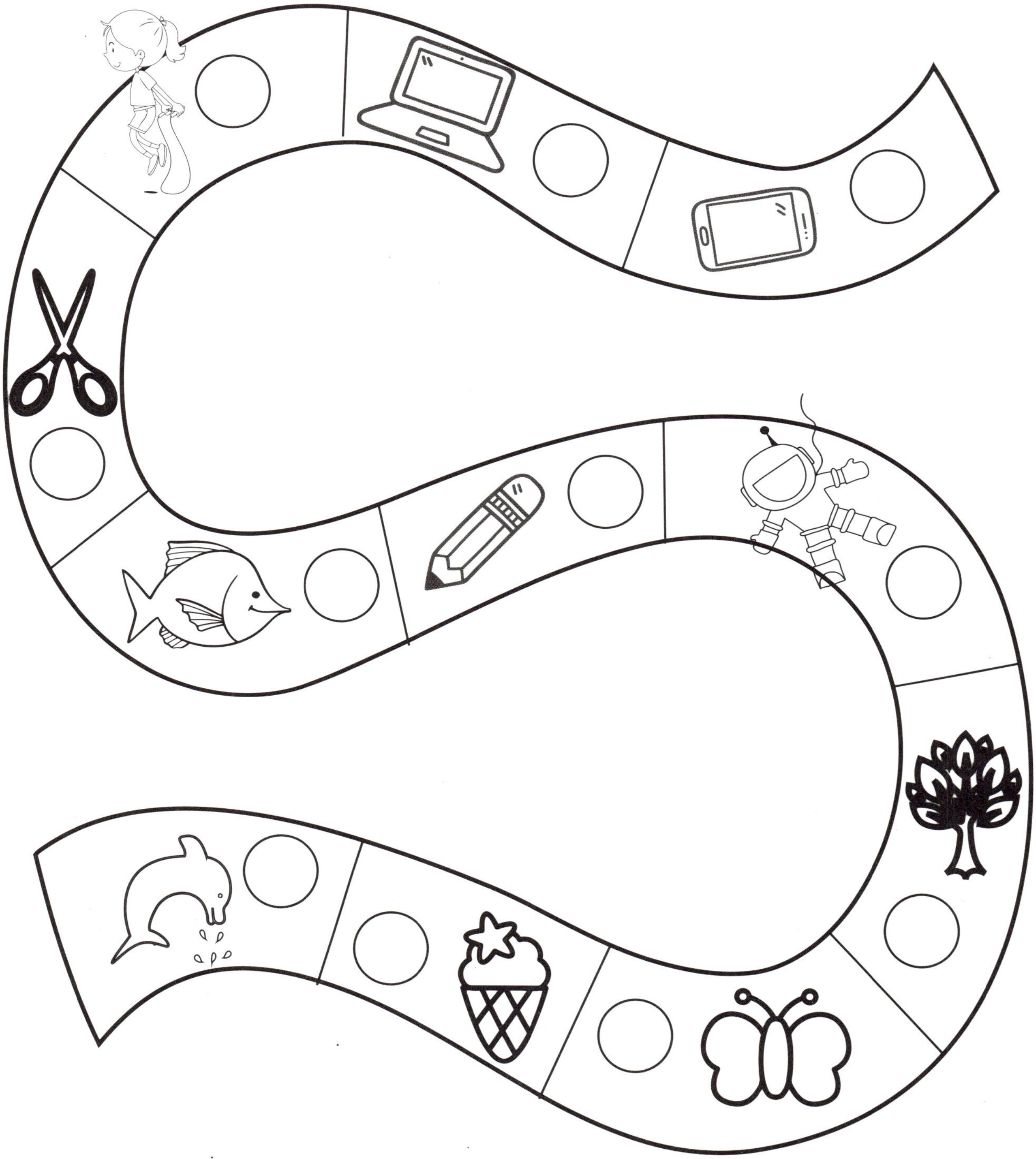

Keep our oceans clean !

For more sheets visit:
www.biplobworld.com

Learn with Biplob

Electrical or Non Electrical

Color the movements that don't require electricity

Learn with Biplob

Push and Pull are Forces

Look at each picture. Is the person using a pushing force or a pulling force?
Write push or pull in the line below each picture.

____

____

____

____

____

____

Food Chain

Grass / plants: PRODUCERS

They are eaten by small insects (eg: grasshopper): PRIMARY CONSUMERS

They are eaten by small birds / animals (eg: frog): SECONDARY CONSUMERS

They are eaten by larger animals (snake), : TERTIARY CONSUMERS

They are eaten by predators (eagle): APEX CONSUMER

They are decomposed by micro – organisms and they go back into the soil: DECOMPOSERS

Categorize the animals in the food chain.

Goat

Lion

Tiger

Lizard

Grass

For more sheets visit:
www.biplobworld.com

Learn with Biplob

Copy the picture

Name the Planet first _______________

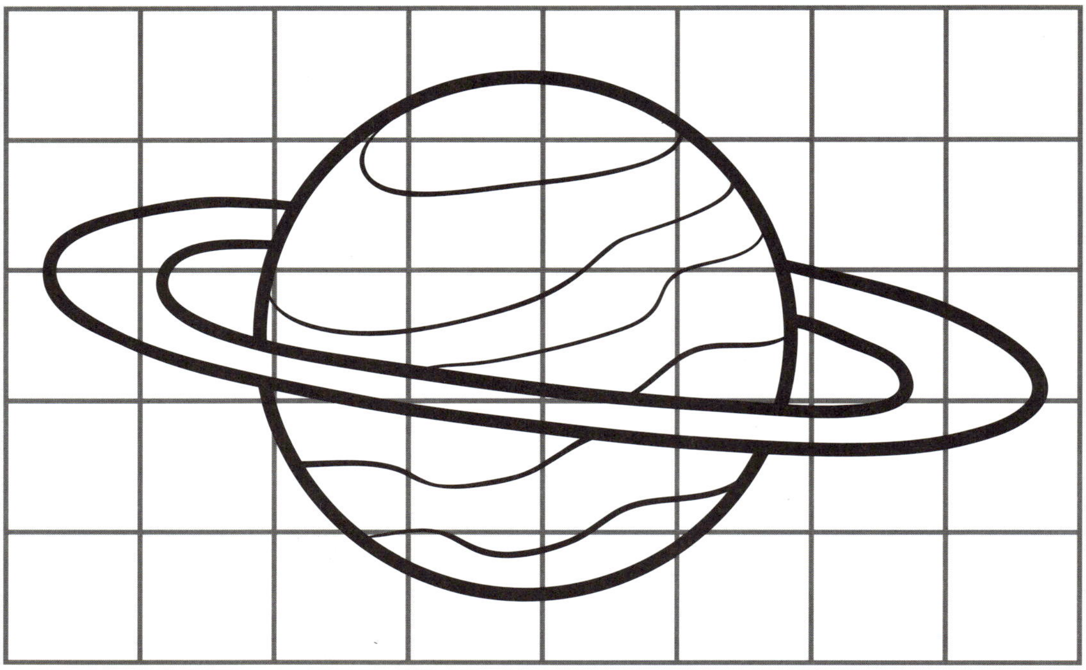

COLOUR ME

with Biplob the Bumblebee

The Adventures of
BIPLOB THE BUMBLEBEE

The Adventures of
BIPLOB THE BUMBLEBEE

The Adventures of
BIPLOB THE BUMBLEBEE

The Adventures of

BIPLOB THE BUMBLEBEE

The Adventures of
BIPLOB THE BUMBLEBEE

The Adventures of
BIPLOB THE BUMBLEBEE

The Adventures of
BIPLOB THE BUMBLEBEE

The Adventures of
BIPLOB THE BUMBLEBEE

The Adventures of
BIPLOB THE BUMBLEBEE

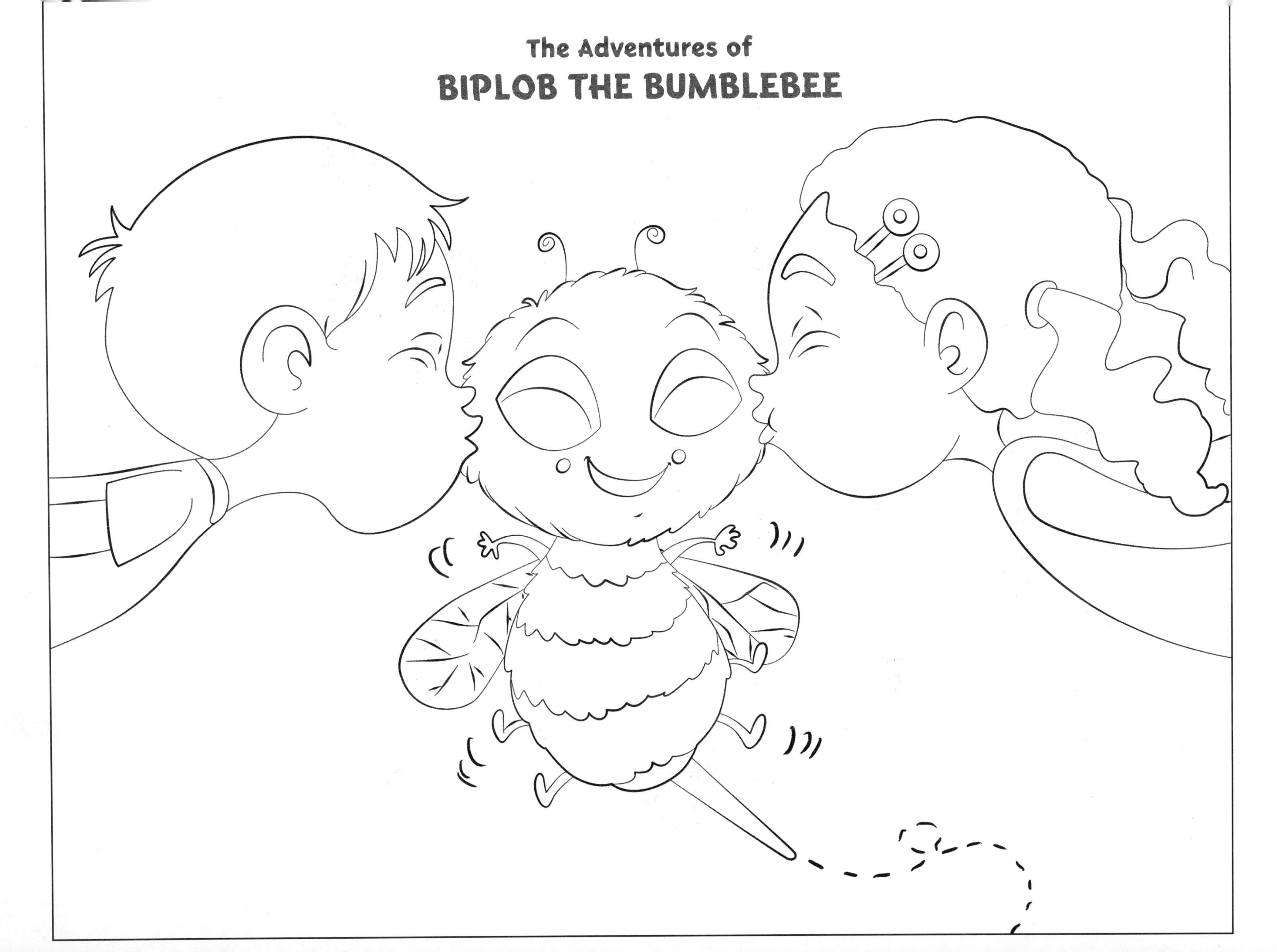

The Adventures of
BIPLOB THE BUMBLEBEE

Happy learning – after all, as Biplob says, learning is the most NATURAL process!

BIPLOB THE BUMBLEBEE
Book Series

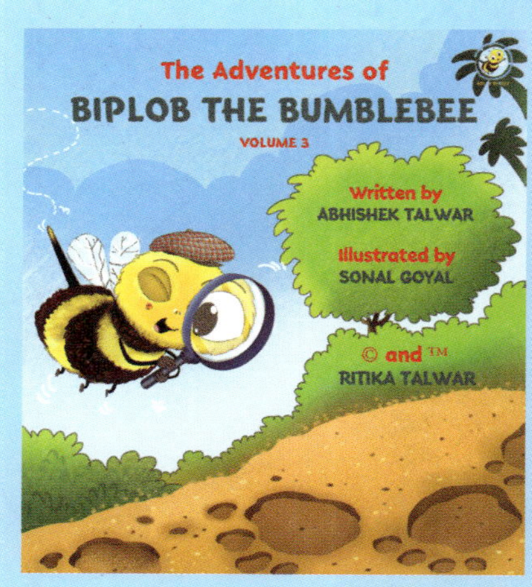

Happy learning – after all, as Biplob says, learning is the most NATURAL process!

BIPLOB THE BUMBLEBEE
Colouring Books and Worksheets